CONTENTS

Electromagnetic Waves	1*
Electromagnetic Waves	1a
Sources of Light	1b
Light Waves	2*
Light Waves	2a
Reflection and Refraction	3*
Reflection and Refraction	3a
Light and Color	4*
Light and Color	4a
Light and Materials	5*
Light and Materials	5a
Lenses	6*
Lenses	6a
Optical Instruments	7*
Optical Instruments	7a
Television	7b
Sound	8*
Sound	8a
Sound Waves	9*
Sound Waves	9a
Characteristics of Sound	10*
Characteristics of Sound	10a
Sound and Materials	10b
Musical Instruments	11*
Musical Instruments	11a
The Human Voice	11b
Development of Communication Systems	12*
Development of Communication Systems	12a

For final testing and review:

A Last Look—Part I	I
A Last Look—Part II	II
A Last Look—Part III	III
A Last Look—Part IV	IV

* Full–color transparencies are found at the back of the book. Each transparency should be used to introduce the corresponding unit.

© Milliken Publishing Company   Light and Sound Energy

TEACHING GUIDE

Page 1 ELECTROMAGNETIC WAVES

CONCEPTS: 1. Light is one type of radiant energy. 2. Light is emitted when electrons move from a higher energy level to a lower one. 3. Objects are luminous, light producers, or illuminated, light reflectors.

BACKGROUND INFORMATION: *Radiant energy* is a kind of energy that travels outward from a central source. Examples include heat and light, both of which are electromagnetic waves. *Electromagnetic waves* have both electric and magnetic components. Scientists classify electromagnetic waves according to their *wavelengths*, the distance between two waves. The range of electromagnetic waves is known as the *electromagnetic spectrum*. For study, the electromagnetic spectrum is divided into eight major regions—cosmic rays, gamma rays, X–rays, ultraviolet rays, visible light *(or the optical spectrum)*, infrared rays, radio waves, and power frequencies *(electricity)*. All of these waves, with the single exception of visible light, are invisible to the naked eye.

Light energy is produced by the activity of electrons around the nucleus of an atom. If an atom absorbs energy, its electrons move to a higher energy level. As these electrons return to the lower energy level, the atom gives off light energy in bundles, which scientists call *photons*. Lifting the electrons back again to a higher level of energy requires exactly the same amount of energy. Movement to a higher level results in a gain of potential energy. The higher the energy level the atom reaches, the more potential energy it gains and the more light it will eventually release.

Light sources are luminous or illuminated. *Luminous* objects produce their own light. Luminous light sources can be either natural or artificial. Our most important source of natural light is the sun. This tremendous mass of incandescent gas, produced by nuclear fusion, provides the earth with light and other types of energy. *Illuminated* objects do not produce light, but are seen because they reflect light from another object. The moon and planets receive light from the sun and reflect it; they are illuminated objects. Fireflies produce light through chemical reactions. Artificial light is made by people. Common sources of this kind of light are candles, kerosene lamps, oil lamps, natural gas lamps, incandescent lamps, fluorescent lights, and neon lights. Electricity is most often used to produce artificial light. However, electricity is generated by water power or steam power from coal. Both power sources can be traced back to the sun.

ENRICHMENT ACTIVITIES: 1. Find out about the light year as a unit of measure for astronomical objects. 2. Research the contributions of Christian Huygens, James Maxwell, Heinrich Hertz, and Max Planck to the understanding of light energy.

ANSWER KEY:
Page 1 1. Light 2. When an electron moves from a higher energy level to a lower one, light energy in the form of photons is released. **Study Question:** The electromagnetic spectrum is the range of electromagnetic waves that includes the various forms of radiant energy. It comprises ranges of different wavelengths in order of size. In 1864, James Maxwell defined visible light as one of several types of electromagnetic waves and predicted that there were other types that were invisible. In 1887, Heinrick Hertz confirmed Maxwell's prediction by demonstrating the existence of wavelengths longer than the wavelength of visible light.
Page 1a 1. A. X–rays B. light rays, ultraviolet rays C. ultraviolet rays D. infrared rays E. radio waves 2. A. I B. V, I C. I D. I E. I 3. a. X–rays b. photons c. electric, magnetic d. electrons, nucleus e. ultraviolet 4. light, heat 5. higher, lower, light

Page 1b 1. moon 2. sun–nuclear fusion, lightning–burning 3. light bulb, neon sign, fluorescent light 4. wood burning and gas burning 5. a. sun b. artificial c. chemical d. sun e. illuminated

Page 2 LIGHT WAVES

CONCEPTS: 1. Light travels away from its source in all directions in a straight line. 2. Light travels at a speed of about 300,000 kilometers (186,300 miles) per second.

BACKGROUND INFORMATION:
Throughout history, scientists have proposed several theories in an attempt to explain the nature of light. Sir Isaac Newton proposed the *particle theory of light,* which considers light to be streams of particles smaller than atoms, moving out in all directions. It was thought that the more particles striking the human eye per second, the brighter the light would appear. The fewer the particles that reach the eye, the dimmer the light would appear. Scientists also theorized that light is waves moving outward in all directions like waves on water. The closer one is to the source, the larger and stronger the waves would be. This would cause the light to appear bright. As the waves move farther from the source, they spread out and grow smaller, and the light appears dim. This theory is called the *wave theory of light.* The present theory of light, developed by Niels Bohr and other scientists, assumes that electrons may move from a higher energy level orbit to a lower energy level orbit. In doing so, small bundles *(quanta)* of energy, called *photons*, are released. The bundles move in waves, which move in straight lines. This is known as the *quantum theory of light.*

Electromagnetic waves move up and down at the same time that they move forward in straight lines. Therefore, each wave has length and frequency. *Wavelength* is the distance between two waves, and *frequency* is the number of waves that pass a given point in one second. A *crest* is when the wavelength reaches the height of its movement; a *trough* is when the wavelength reaches the depth of its movement. Light is not instantaneous, but it is extremely fast, moving about 300,000 kilometers (186,300 miles) per second.

ENRICHMENT ACTIVITIES: 1. Research the difference of the speed of light in air and the speed of light in a vacuum. 2. Find out about the measure of intensity of light.

ANSWER KEY:
Page 2 1. According to the quantum theory, electrons may move from a higher energy level orbit to a lower energy level orbit. When the electrons move, bundles of light energy, called quanta, are released. 2. Light travels in waves and in a straight line.
Study Question: The optical or light spectrum includes the visible part of the electromagnetic spectrum. It is a very small part of the entire spectrum and includes those wavelengths or radiant energy that are visible to the human eye.
Page 2a 1. A. particle B. wave C. quantum 2. a. wavelength b. crest c. trough 3. A bundle of light energy (quantum) travels from the atom in waves. 4. wave theory of light 5. a. trough b. frequency c. wavelength d. crest 6. 300,000, second 7. a straight line

Page 3 REFLECTION AND REFRACTION

CONCEPTS: 1. Light rays are reflected when they bounce off an object or substance. 2. Light rays are refracted, or bent, as they pass obliquely from one substance into another that has a different density.

BACKGROUND INFORMATION: Just as a ball will bounce off a surface, such as a wall or a person's hands, light rays bounce off of some objects. This is known as *reflection*. The light ray that strikes a surface is called the *incidence ray*, and the ray that is reflected is called the *reflected ray*. The angle between the light ray that strikes the surface at a point and a perpendicular line at that point is called the *angle of incidence*. The angle between the reflected ray and the perpendicular line is called the *angle of reflection*. The law of reflection states that the angle of incidence is equal to the angle of reflection. This law can be applied to smooth, polished surfaces, but not to rough surfaces. Rough surfaces scatter or spread light rays in all directions at mixed angles, producing poor or no reflections.

Unlike reflected rays, refracted rays pass through the surface of the object they hit. A perpendicular light ray traveling through a transparent *(clear)* material travels in a straight line and continues at the same speed. When the light ray passes at a slant from one transparent material into another transparent material *(such as air into water)*, the light ray travels in a different direction and at a slower speed. This causes the light ray to be *refracted*, or to bend. Refraction of light rays only occurs when the rays pass through the transparent material at a slant; refraction does not occur when light travels in a perpendicular line. The change in the speed of the light rays is a cause of refraction. Differences in speed depend on the density of the material. The greater the density, the slower the speed and the greater the refraction. Water is more dense than air. As light rays pass through the air and strike the water at a slant, they slow down, refract, and often deceive our vision.

ENRICHMENT ACTIVITIES: 1. Research the reflections caused by curved mirrors. 2. Find out how refraction affects spearing fish under water.

ANSWER KEY:
Page 3 1. Light is scattered, and few of the rays reach the eye. 2. Refraction of light is caused by the bending of light rays and a change in their speed as they move from one substance to another. **Study Question:** Mirages are optical illusions caused by the refraction of light as it passes through layers of air having different densities due to heating.
Page 3a 1. A. incidence ray B. angle of incidence C. reflected ray D. angle of reflection 2. light and a smooth, polished surface 3. smooth surface 4. reflection 5. slows down 6. Water is more dense than air. As light rays pass through the air and strike the water at a slant, they slow down and refract.

Page 4 LIGHT AND COLOR

CONCEPTS: 1. A beam of white light passing through a prism forms a spectrum of color. 2. The order of the colors of the spectrum from the longest wavelength to the shortest are red, orange, yellow, green, blue, indigo, and violet.

BACKGROUND INFORMATION: *White light* is really a mixture of many colors. These colors can be seen when a beam of sunlight passes at a slant through a glass *prism*. The prism breaks up the white light into bands of colored light called a *spectrum*. The seven colored lights in the spectrum are violet, indigo, blue, green, yellow, orange, and red. Colors of the spectrum are seen in that order because our eyes see each wavelength as a different color. The prism refracts the colored lights in varying amounts; short waves are bent more than long waves. Violet light, with the shortest wavelength, is bent the most; red light, with the longest wavelength, is bent the least.

Blue–colored material looks blue because the chemical structure of that material absorbs all the colored waves except blue, which is reflected and seen by the eye. White–colored material appears white because it reflects all the colored waves. Black–colored material appears black because it absorbs all the colored waves and reflects none. Dark colors absorb light energy and change it into heat. Therefore, dark clothes worn in winter help to keep us warmer. Light colors worn in summer help to keep us cooler because these colors reflect more light, absorbing less energy.

ENRICHMENT ACTIVITIES: 1. Learn how a rainbow is formed in the sky. 2. Find out about the additive and subtractive combinations of primary colors.

ANSWER KEY:
Page 4 1. Red, orange, yellow, green, blue, indigo, violet. 2. The rose is seen as red because all the colored rays except red are absorbed and therefore only the red ray is reflected to our eyes. **Study Question:** Outer space is black because there is nothing to reflect light.
Page 4a 1. A. violet B. indigo C. blue D. green E. yellow F. orange G. red 2. Answers will vary. 3. a. violet b. red c. rainbow d. blue 4. The prism breaks up the white light into bands of colored light. 5. The darker color absorbs more heat and reflects less than the roof of the white car.

Page 5 LIGHT AND MATERIALS

CONCEPTS: 1. Light rays can pass through some materials. 2. An object that blocks light rays produces a shadow.

BACKGROUND INFORMATION: Materials such as clear glass and plastic, air, and water let almost all light rays pass through them. These are called *transparent materials,* and we can see clearly through them. Materials such as oiled paper, frosted glass, and some plastics allow some light rays to pass through them. These are called *translucent materials.* We cannot see clearly through translucent materials because the light rays are refracted. They change direction and scatter (diffuse) as they pass through the translucent materials. Materials such as wood, clay, brick, and metals allow no light rays to pass through them. These materials reflect all light and are called *opaque.* Most of the light rays striking opaque materials are blocked. Therefore, we cannot see through them.

Any time an opaque material blocks the passage of light, shadows form. If the source of light has a surface area and is not just a point of light, the opaque object casts an entirely dark shadow, which scientists call the *umbra,* and a partial shadow, called the *penumbra,* where the light is not completely blocked. The penumbra is absent if the light source is just a point, rather than an area.

ENRICHMENT ACTIVITIES: 1. Research Polaroid lenses and their ability to absorb light. 2. Find out how to make a shadow grow larger.

ANSWER KEY:
Page 5 1. transparent 2. A shadow is formed when an opaque object blocks the passage of light. **Study Question:** When an opaque object blocks an extended source of light, a full shadow (umbra) and a partial shadow (penumbra) are produced.
Page 5a 1. A. transparent B. translucent C. opaque 2. a. light passes through b. light is scattered c. no light passes through 3. a. TP or TL b. TL c. TL d. TP e. OP f. OP g. OP h. TP i. TL 4. a. opaque object b. transparent object c. shadow d. translucent object 5. When an opaque material blocks the passage of light, an entirely dark shadow (umbra) and a partial shadow (penumbra) are produced.

Page 6 LENSES

CONCEPTS: 1. A lens is a curved piece of transparent material used to refract light rays. 2. The two most common kinds of lenses are convex and concave lenses.

BACKGROUND INFORMATION: A curved glass lens refracts light rays that pass through it. When light rays pass through a lens, they always bend toward the thickest part of the lens. The most common kinds of lenses are convex and concave lenses. *Convex lenses* are thick in the middle and thin at the ends. They can produce an image by bending the light rays and focusing them. The image, however, appears upside down. The closer the object is to the lens, the larger the inverted image is. The farther an object is from the lens, the smaller the inverted image is. *Concave lenses* are thin in the middle and thick at the ends. Light rays passing through concave lenses spread apart instead of meeting at the point. The images produced by concave lenses are smaller than the object and right side up.

The lenses of our eyes work in the same way. Light rays passing through the lens of the eye produce an image which is focused on the *retina*. The curvature of the eye's convex lens is changed by muscles attached to it. This changing enables the eye to focus on near or distant objects. If the lens focuses the image exactly upon the retina, the eyesight is normal. When the eyeball is elongated or the lens is too convex, the image is in focus in front of the retina. This condition is called *nearsightedness*. People having this condition can see nearby things very well, but not distant things. A nearsighted condition can be corrected by concave lenses, which move the focusing point back to the retina. When the lens is too concave or the eyeball is not elongated enough, the image is in focus behind the retina. This condition is called *farsightedness*. People having this condition see things well at a distance but not nearby. This condition can be corrected by convex lenses, which bring the focus forward to the retina. Farsightedness is a common eye defect of middle–aged and older persons, because their eye muscles lose their power to adjust the eye.

ENRICHMENT ACTIVITIES: 1. Find out how glass lenses are ground and polished. 2. Find out about the defects of lenses called chromatic aberration and astigmatism.

ANSWER KEY:
Page 6 1. Light passing through a lens is refracted. 2. The human eye has a convex lens. **Study Question:** Concave lenses correct nearsightedness by moving the focusing point back to the retina. Convex lenses correct farsightedness by bringing the focus forward to the retina.
Page 6a 1. A lens is a curved piece of transparent material used to refract light rays. 2. A. convex B. concave 3. a. concave, convex b. convex c. muscles d. near e. far 4. upside down 5. It is the focal point where the rays meet. 6. They spread out.

Page 7 OPTICAL INSTRUMENTS

CONCEPT: Optical instruments use lenses to refract, focus, and/or magnify images.

BACKGROUND INFORMATION: Common examples of *optical instruments* are the camera, telescope, motion picture projector, microscope, eyeglasses, and binoculars. Convex lenses, rather than concave, are used in most optical instruments. The lens of a camera refracts light rays and focuses the image upon the film, producing a negative print of the image. A *refracting telescope* has two lenses. At one end, the lens called the *objective* gathers and refracts light rays. At the other end, near the eye, the lens called the *eyepiece* magnifies the image produced by the objective lens. To focus a camera or a telescope, the lens is moved back and forth. The lens in a movie projector refracts the light passing through the camera and produces a large image on the screen. A *microscope* has two lenses and works very much like a refracting telescope. Its *objective lens* gathers the light and refracts it to produce an image. The convex lens in the *eyepiece* magnifies the image produced by the larger objective lens. To focus a microscope, the part of the tube housing the eyepiece lens slides toward or away from the objective lens.

Television is the communication of moving pictures and sound between distant places. A *camera lens* focuses the image onto a light sensitive screen. Electrons are released from the other side of the screen and drawn to an electrically charged target plate. An image of dark and light patterns is formed on the plate. An *electron gun* scans the target plate very rapidly with a beam of electrons. A *collector* gathers up reflected electrons. With audio signals, the video signals are relayed to a *transmitter* that combines them with a carrier wave to be sent out in all directions. Signals can also be sent through cables. The *antenna* on the television receiver at home picks up the signals. The video signals are converted to an image on the picture tube when the receiver's electron gun scans the special chemical coating on the screen. The audio signals are converted to sound waves we can hear.

ENRICHMENT ACTIVITIES: 1. Research the lens arrangement in reflecting telescopes like the one on Mount Palomar. 2. Find out about the work of Leeuwenhock and Galileo with lenses.

ANSWER KEY:
Page 7 1. convex 2. One lens gathers, refracts, and focuses the light, and the other lens magnifies the images. **Study Question:** Binoculars are optical instruments that have lenses like telescopes, but also use reflecting prisms to reduce the length of each barrel of the instrument.
Page 7a 1. The lens (objective) refracts the light. 2. convex 3. Use illustrations on page 7 to determine direction of light rays.
4. a. microscope b. refracting telescope c. camera d. movie projector 5. Focusing is done by moving a lens. 6. The camera. Light is focused by a convex lens onto a light sensitive structure (film or retina).
Page 7b 1. Camera, Transmitter, Receiver 2. a. electron gun b. lens c. collector d. transmitter e. antenna 3. strengthens the video or audio signal 4. The chemical coating on the screen is caused to glow by the electrons being directed to the screen from the electron gun. 5. To gather in the signals being relayed from a satellite in the sky.

Page 8 SOUND

CONCEPTS: 1. Sounds are caused by objects vibrating. 2. Sound waves are received by the ear and interpreted by the brain.

BACKGROUND INFORMATION: *Sound* is a vibratory disturbance in a substance or a medium such as air. Sounds occur when things *vibrate*. Vibrating objects send out sound waves that travel in all directions. *Sound waves* can travel through solids, liquids, and gases by causing these substances to vibrate. However, sound waves cannot move through a vacuum because there is no medium present to vibrate. Sound waves travel at different rates through different substances. As they travel, sound waves can also cause nearby substances to vibrate. This helps to propagate the sound wave. When sound waves strike the ear, they are channelled through the ear canal and strike the eardrum. The vibrations of the eardrum cause the minute bones of the inner ear to vibrate. The vibration of these bones causes the fluid that stimulates hearing receptors in the cochlea to vibrate. The auditory nerve then relays the sensation to the brain, which interprets the stimulus. Through experience we learn to associate certain sounds with certain events. This allows us to recognize the voice of a friend, as well as to distinguish it from the sound of glass breaking.

ENRICHMENT ACTIVITIES: 1. Learn how sound waves have been photographed. 2. Find out about "whispering galleries."

ANSWER KEY:
Page 8 1. Sound is caused by vibrations. 2. Vibrations of air molecules are received by the ear and cause the eardrum to vibrate. The stimulus is then transmitted to the brain. **Study Question:** The moon has no atmosphere—there is no medium in which sound waves can travel—and therefore no sound is transmitted.
Page 8a 1. tines of tuning fork, vocal cords, strings of violin, drumhead, bell of doorbell, air 2. a. cup–like structure of bell b. speaker c. strings d. hanging pieces e. vocal cords f. strings 3. a. vibrations b. vibrating c. sound waves, vacuum d. bones e. auditory 4. When the bell rings, vibrations are sent through the air. The vibrations are then received by the ear and transmitted to the brain, where the stimulus is interpreted. 5. experience

Page 9 SOUND WAVES

CONCEPTS: 1. A sound wave is a series of compression and rarefaction waves that spread out in all directions from the sound source. 2. Sound waves travel at a speed of 331.5 meters (361.3 yards) per second at 0° C (32° F).

BACKGROUND INFORMATION: Any object or medium caused to vibrate produces *sound waves*. In air, the forward movement of vibrating objects pushes molecules together. This is called *compression*. When the vibrating object moves back in the opposite direction, the air is separated, causing the molecules to move farther apart. This is called *rarefaction*. A sound is produced by matter when a succession of compression and rarefaction disturbances occurs which can be heard by the human ear or detected by an instrument. Like light waves, sound waves spread out in all directions from the source. Sound waves, however, travel much slower than light waves. They travel through different substances at different speeds. Due to the closeness of the molecules in the solid and liquid states of matter, sound waves generally travel faster in solids and liquids than in gases. The speed of sound in a piece of steel is more than 15 times faster than it is in air at the same temperature (25°C—77° F). Temperature also has an effect on the speed of sound waves. It has a relatively small effect in solids and liquids, but in gases, the speed of sound increases at a rate of 0.6 meters (2 feet) per second for each increase in degree Celsius.

ENRICHMENT ACTIVITIES: 1. Learn how sonar locates objects under water. 2. Find out how Native Americans were able to track the movements of buffalo herds without being able to see them.

ANSWER KEY:
Page 9 1. A series of compression and rarefaction waves move through matter, producing a sound. 2. Through a solid. **Study Question:** An echo is produced when a sound wave is reflected from a hard surface that is at least 16.5 meters (18 yards) distant.
Page 9a 1. A. wavelength B. trough C. crest 2. compression—molecules close together, rarefaction—molecules farther apart 3. Individual molecules do not travel, but vibrate rhythmically back and forth. 4. a. 331.5 m/sec. b. 343.5 m/sec. c. 346.5 m/sec. d. 349.5 m/sec. e. 354.3 m/sec. f. 356.7 m/sec. g. 365.1 m/sec. h. 372.3 m/sec. 5. a. liquid b. gas c. solid d. solid

Page 10 CHARACTERISTICS OF SOUND

CONCEPTS: 1. Frequency is the number of vibrations an object makes per second. 2. Amplitude is the distance a vibrating particle has been displaced from its resting position.

BACKGROUND INFORMATION: Sound waves have certain characteristics which enable us to distinguish one sound from another. The *frequency* of a sound wave is the number of waves that pass a certain point per unit of time. A rapidly vibrating object has a high frequency. High frequency waves produce a high–pitched sound. Slowly vibrating objects have a low frequency and produce a low–pitched sound. The human ear has a normal range capable of hearing sounds about as low as 16 vibrations per second (vps) and as high as 16,000 vibrations per second. Sounds below 16 vps are called *subsonic,* and those above 16,000 vps are known as *ultrasonic.* Ultrasonic sound waves are used in various industrial and medical activities. *Amplitude* is the distance a vibrating particle has been displaced from a resting position. Amplitude determines the intensity of a sound. *Intensity* is defined as the amount of energy flowing in the sound waves. The larger the amplitude, the more intense the sound. *Loudness* is how strong the sound seems when it reaches our ears. Although related to intensity, it is not the same concept due to the interaction of frequency. If frequency is held constant, loudness and intensity vary together. That means that as one increases the other increases at the same rate. However, equally intense sounds traveling at different frequencies are not equally loud. When intensity is held constant, sounds in the middle ranges of frequency are louder than sounds in the upper or lower frequency ranges. *Loudness* is measured in decibel units by sound meters. Sounds above 120 decibels are painful to the ears and can cause injury to the ear if prolonged.

Some substances conduct sound better than others do. Most sounds come to us through air, a gas. Gases are not the best conductors of sound. The molecules in a gas are spaced far apart, making it a poor conductor of sound energy. Where the air is thin and cool, such as in high altitudes, sound does not travel as fast or as far as compared to lower altitudes where the air is warmer and denser. Because the molecules are closer together in a liquid than in air, liquids conduct sound better and faster than gases. Sound travels best in hard solids, where molecules are very close together. When a sound wave bounces back *(reflects)* from a large hard surface, it may be heard as an echo. To hear the echo, a person must be at least 16.5 meters (18 yards) away from the reflecting source. Annoying echoes can be eliminated by using *acoustical* material to absorb the sound. Concert halls and gymnasiums are built with this principle in mind.

ENRICHMENT ACTIVITIES: 1. Research the use of ultrasound in physical therapy. 2. Find out about the quality of sounds that enable us to distinguish the voices of friends.

ANSWER KEY:
Page 10 1. A high pitched sound would be produced. 2. No, it would be above the normal limits of the human ear. **Study Question:** The Doppler effect is heard when a sound–producing object approaches the listener, and the sound waves are compacted, causing an increase in pitch. As the object moves away, the sound waves are stretched out, and a decrease in pitch is heard. A fire engine siren coming and going displays the Doppler effect.
Page 10a 1. I. Frequency is the number of vibrations per second. II. Amplitude is the distance a vibrating particle has been displaced from a resting position. III. Intensity is the amount of energy flowing in the sound waves. IV. Loudness is how strong the sound seems when it reaches our ears. 2. a. high b. low c. low d. high e. high f. low 3. a. high, low b. 16,000 c. subsonic d. frequency e. 120 4. a. 120 b. 50–60 c. 110–120 d. 10–30
Page 10b 1. molecules 2. The friend with

© Milliken Publishing Company   Light and Sound Energy

his or her head underwater. 3. They put their ears onto the track (do not attempt this; it can be dangerous). Sounds travel faster through solids than through air (gas). 4. Materials are soft, irregular, and have holes. 5. There is no air to transmit the sound waves.

Page 11 MUSICAL INSTRUMENTS

CONCEPT: The three basic types of musical instruments are string, wind, and percussion.

BACKGROUND INFORMATION: *String instruments* produce sounds with one or more vibrating strings. The strings vibrate when the musician plucks them or draws a bow across them. Higher pitch can be attained by tightening or shortening the strings, or by using thinner strings. Opposite actions will produce a lower pitch. The intensity of a string instrument sound depends on how hard the strings are plucked or rubbed.

Wind instruments depend upon the vibration of a column of air to produce sound. The column of air vibrates when the musician blows into or across the instrument. Brass and woodwind instruments make up the two types of wind instruments. *Brass instruments* are played by vibrating the lips, which are pressed against the mouthpiece of the instrument. This causes the air column to vibrate. *Woodwind instruments*, such as clarinets, need a reed to make the air column vibrate. The column of air vibrates in the flute and piccolo when air is blown across a hole. Higher pitch can be produced in brass and woodwinds by shortening the column of air. The intensity of wind instrument sound depends on how hard air is blown into the instrument.

Most *percussion instruments* produce sounds when the material stretched over a hollow container vibrates when it is struck with a stick or mallet. Some percussion instruments are solid and vibrate when struck. Higher pitch can be produced by tightening the stretched material, or by using a thinner or smaller piece of material. The intensity of sound depends on how hard the instrument is struck.

The human voice is a very powerful instrument. *Vocal cords* are strong bands of tissue stretched over the top of the voice box *(larynx)*. A narrow slit called the *glottis* separates the cords. Sound is produced when air from the lungs is blown through the glottis, which causes the vocal cords to vibrate. Voice pitch is controlled by muscles attached to the vocal cords which can make the cords tight or loose. Tight cords vibrate rapidly, producing high-pitched sounds; loose cords vibrate slowly, producing low-pitched sounds. Controlling the tension of these muscles to adjust the pitch of sound is something we learn as we are learning to speak. Other areas that affect the quality of the voice are lips, tongue, teeth, and head sinuses.

ENRICHMENT ACTIVITIES: 1. Find out how musical instruments are constructed to have more resonance. 2. Find out why orchestras must tune their instruments several times during a performance.

ANSWER KEY:
Page 11 1. Pitch is changed by tightening or loosening the string, shortening or lengthening the string, or by using thick or thin strings. 2. The membrane on a bass drum is looser, thicker, and larger; it also has a larger resonating chamber. **Study Question:** oboe—wind; harp—string; celesta—percussion; glockenspiel—percussion; zither—string
Page 11a 1. cello—strings, a bow; clarinet—air column, reed 2. tighten string, shorten string, or use a thinner string 3. higher pitch—shorten air column, lower—lengthen air column 4. triangle—use different material, change length of materials;

drum—tighten or loosen covering, use thick or thin covering material, use large or small piece for covering 5. The keys are struck with the fingers, which cause hammers to strike the strings. 6. The fingers cause the strings to vibrate.

Page 11b 1. glottis, air 2. a.V. b.NV c.V d. NV e. NV f. V or NV 3. Muscles can make vocal cords tighter or looser, which effects pitch. 4. a. windpipe b. because it is an air column through which air (wind) passes 5. high, low

Page 12 DEVELOPMENT OF COMMUNICATION SYSTEMS

CONCEPTS: 1. Auditory and visual signals have been used by people to communicate since ancient times. 2. Modern methods of communication enable people to communicate worldwide almost instantaneously.

BACKGROUND INFORMATION: Without the development of a means of communication, people might never have advanced beyond the first stages of social development. Pointing and using guttural sounds were probably the first human means of communicating. The development of speech and language enabled people to transmit knowledge and cultural heritage. Communication over long distances was an important discovery because it allowed people to convey information to one another by indirect means. Smoke signals and drumbeats may have been the first media used to transmit messages over long distances. With the discovery of electrical energy and the means to control it, the capability of communication devices increased greatly. The invention of the telegraph, telephone, radio, and television have made communication rapid and far–reaching. The telephone usually uses wires to send electrical energy into a specific receiver, while technologies such as the radio or television send waves into the atmosphere to be picked up by any of a number of receivers. Modern audio and visual communication systems include the use of satellites that are in synchronous orbit around the earth. Events happening on one part of the earth can be seen and heard on any other part of the globe almost instantly.

ENRICHMENT ACTIVITIES: 1. Research the developments that lead to the discovery of radio waves. 2. Find out about the impact of transistors and printed circuits on communication devices.

ANSWER KEY:
Page 12 1. electricity 2. Most telephone signals are transmitted through wires. Radio signals are transmitted by radio waves in the atmosphere. **Study Question:** Animals use a variety of methods to communicate with their own and other species. These include sounds, gestures, colors, and odors.
Page 12a 1. sender, receiver 2. Drum signals could be used in wet and dry weather and during the day or at night. 3. To use a telegraph, a person had to become well versed in telegraph code (Morse Code) in order to send and receive messages, and messages could be sent only where wires were strung. 4. audio, video 5. sound, electrical, sound 6. telephone, radio, television.

A LAST LOOK—PART I

A. 1. X–rays do not belong. They are electromagnetic waves, while radio and sound waves carry sound.
   2. Compression does not belong. It refers to molecules and sound. Refraction and reflection refer to materials and light.
   3. Moonlight does not belong. It is illuminated light, not luminous light like firelight and sunlight.
   4. Black does not belong. It is the absence of color, while red and blue are both reflections of color.

5. Screen does not belong. It is not a type of lens, as are convex and concave.
6. Radio does not belong. It is not an optical instrument containing lenses.
7. Photon does not belong. It is the name for light energy. Crest and trough are parts of a wave.
8. Telephone does not belong. It is a communications device, not a musical instrument.
9. Pitch does not belong. It refers to sound, while particle and quantum are theories of light.
10. Rarefaction does not belong. It is part of a sound wave. Amplitude and frequency are measurements of a sound wave.

Note: These are suggested answers, each determined by a specific viewpoint. Since more than one correct answer is possible, accept any reasonable answer a student can justify.

B. 1. radiant   6. Opaque
   2. light   7. vibrations
   3. crest   8. solid or liquid
   4. 300,000 km/sec 9. ultrasonic
   5. prism   10. eardrums

Note: If the ability of the students might prohibit them from writing the words required to complete Part B of this page, the answers might be randomly displayed on the chalkboard or bulletin board.

A LAST LOOK—PART II

A. 1. a. microscope   b. convex
   2. a. camera   b. convex
   3. a. telescope   b. convex
   4. a. movie projector b. convex

B. 1. a. violin b. tightening/loosening or shortening/lengthening strings
   2. a. saxophone b. shortening/lengthening air column
   3. a. guitar b. tightening/loosening or shortening/lengthening strings
   4. a. drum b. stretch material on top of drum, or use thinner material

C. 1. a. fire   b. luminous
   2. a. lightbulb b. luminous
   3. a. moon   b. illuminated
   4. a. candle   b. luminous

A LAST LOOK—PART III

A. 1. l   4. h   7. f   10. e
   2. a   5. k   8. b
   3. d   6. i   9. g

B. 1. television   6. sympathetic
   2. vocal cords   7. Convex
   3. pitch   8. opaque
   4. gases   9. white
   5. decibels   10. slow down

A LAST LOOK—PART IV

A. Radiant energy is any kind of energy that travels outward towards a central source, such as light or heat. Radiant energy is not found buried in the ground.

B. 1. The order of colors in the spectrum is violet, indigo, blue, green, yellow, orange, and red.
   2. The boy is playing a string instrument, not a wind instrument.
   3. The trough should be at the bottom of the wave and the crest at the top.
   4. Trees, not a large, hard surface, would not produce an echo.

C. 1. photon   9. reflection
   2. quantum   10. decibel
   3. sound   11. opaque
   4. prism   12. camera
   5. infrared   13. pitch
   6. colors   14. waves
   7. shadow   15. light
   8. drum   16. concave
   shaded boxes: rarefaction

Electromagnetic Waves

Kinds of Electromagnetic Waves

A. _____   B. _____

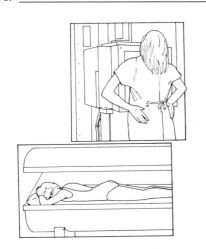
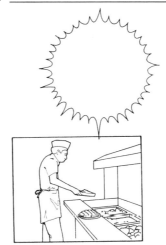
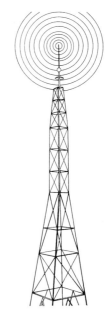

C. _____   D. _____   E. _____

Photons of Light

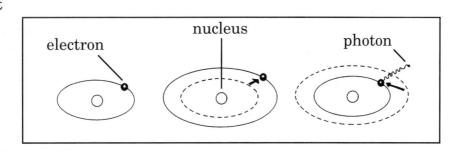

1. Label the kinds of electromagnetic waves on lines A through E above.
2. Behind each label, write V or I to indicate whether that type of electromagnetic wave is visible or invisible.
3. Write the word or words that will make each sentence a true statement.

   a. _____ are used to detect broken bones.

   b. Bundles of light energy produced by electron movement are called _____ .

   c. Electromagnetic waves have both _____ and _____ components.

   d. Light energy is produced by the activity of _____ around the _____ of an atom.

   e. People use _____ rays to get a tan.

4. List two examples of radiant energy: _____.
5. Electrons that move to a *(higher, lower)* level gain energy; electrons that move to a *(higher, lower)* level give off _____ energy.

© Milliken Publishing Company

Sources of Light

Luminous objects produce light. Illuminated objects reflect light.

Natural Light Sources

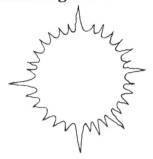

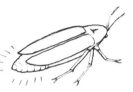
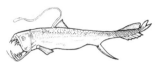

sun—nuclear fusion   lightning—burning   moon—reflected   firefly—chemical   deep-sea fish—chemical

Light Sources

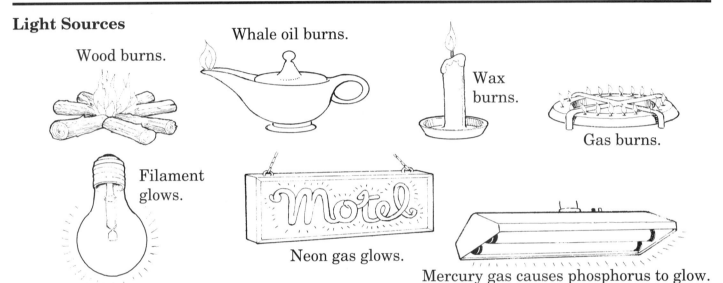

1. Circle the picture of a natural light source that is not luminous.
2. Which sources of natural light also produce heat? _____
3. Circle three artificial light sources that require electrical energy to produce light.
4. Which two artificial sources of light pictured here are most often used for heat?
   _____
5. Write the word or words that will make each sentence a true statement.
   a. Our most important source of light energy is the _____.
   b. We are able to see objects around us because of _____ light.
   c. Deep-sea fish and fireflies produce light through a _____ reaction.
   d. Even though electricity is generated by water or steam, it can still be traced back to the _____.
   e. Planets are _____ objects because they reflect light from another source.

Light Waves

Theories of Light

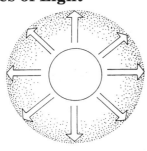 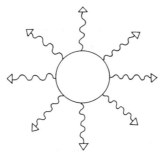 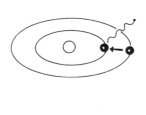

A. _____ B. _____ C. _____

Parts of a Light Wave

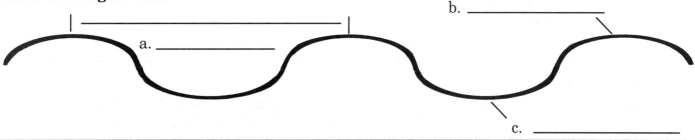

Speed and Movement of Light

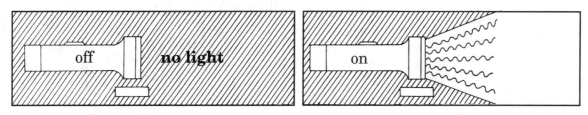

1. Write the names of the theories of light on lines A through C above.
2. Label the parts of a light wave on lines a through c above.
3. Briefly explain the most modern theory of light. _____
   _____
4. Which theory compares light movement to dropping a pebble into a pond? _____
   _____
5. Write the word from Column B in the space before its description in Column A.

	Column A	Column B
_____	a. the lowest depth of a wavelength	wavelength
_____	b. the number of waves per second	crest
_____	c. the distance between two waves	trough
_____	d. the highest point of a wave	frequency

6. The speed of light is about _____ kilometers per _____.
7. Light waves travel in _____ as they move away from the source.

© Milliken Publishing Company — Light and Sound Energy

Reflection and Refraction

Reflection of Light

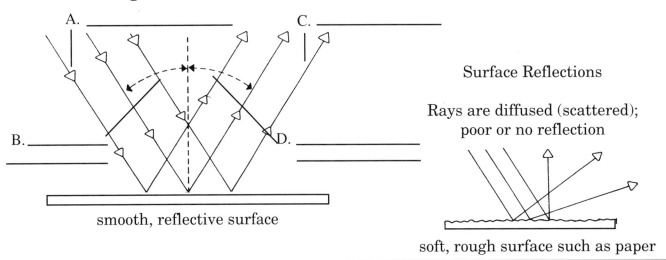

Refraction of Light

Refraction is the bending of light rays.

Light rays bend as they pass through a glass lens.

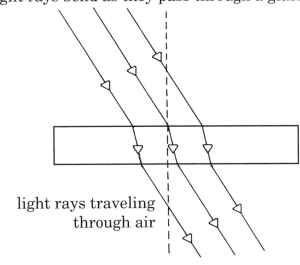

Density and Refraction
Rays slow down in glass.
Glass is denser than air.

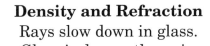

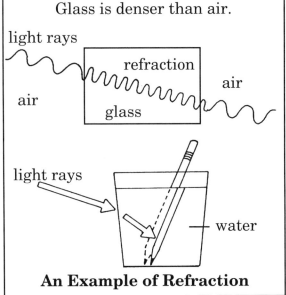

An Example of Refraction

1. Label the parts of reflected light on lines A through D above.
2. What two things are necessary to produce a reflection? _____
3. Light rays are reflected from the _____ of an object.
4. The law of _____ can be applied to smooth surfaces but not to rough ones.
5. A light ray *(speeds up, slows down)* when it passes into a denser material.
6. Why do objects that lie half in and half out of water appear distorted? _____

Light and Color

Colors That Make Up White Light

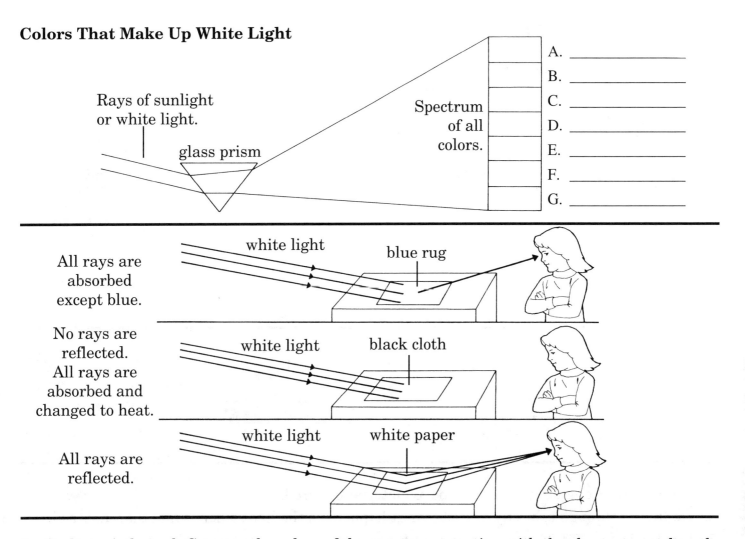

1. On lines A through G, name the colors of the spectrum, starting with the shortest wavelength and ending with the longest.
2. Use the first letter of each color, in ascending or descending order of the spectrum, to create a phrase that will help you to remember the correct order: for example, "Ralph only yelled 'Gosh!' because I veered."

___

3. Write the word or words that will make each sentence a true statement.

    a. Short waves, such as the color _____ , are bent more than long waves.

    b. The color _____ has the longest wavelength.

    c. Another name often used for the spectrum is _____ .

    d. If a rug absorbs all rays except _____ , it will appear blue.

4. Why does the spectrum appear as a band of colors?

___

5. Why is a brown car in the sunshine hotter to the touch than a white car?

Light and Materials

Kinds of Materials and Light

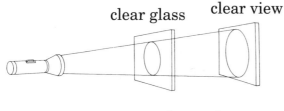

A. _____
a. _____

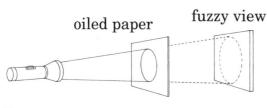

B. _____
b. _____

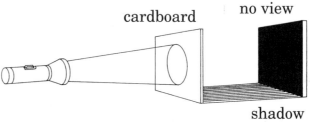

C. _____
c. _____

Light and Shadows

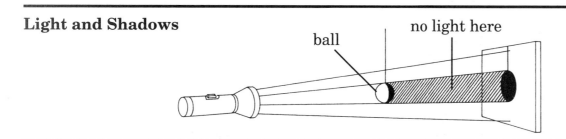

1. On lines A through C, name the type of material pictured.
2. On lines a through c, briefly describe what happens to the light when it hits that material.
3. Use the letters TP, TL, and OP to identify these materials as transparent, translucent, or opaque.

   a. water _____   d. air _____   g. concrete _____

   b. frosted glass _____   e. aluminum foil _____   h. plastic wrap _____

   c. wax paper _____   f. orange juice _____   i. tissue paper _____

4. Write the word or phrase from Column B in the space before its description in Column A.

Column A		Column B
_____	a. produces a shadow	transparent object
_____	b. transmits light rays	opaque object
_____	c. area of darkness	translucent object
_____	d. scatters light rays	shadow

5. What is a shadow? _____

_____

© Milliken Publishing Company   Light and Sound Energy

Lenses

Shapes of Lenses

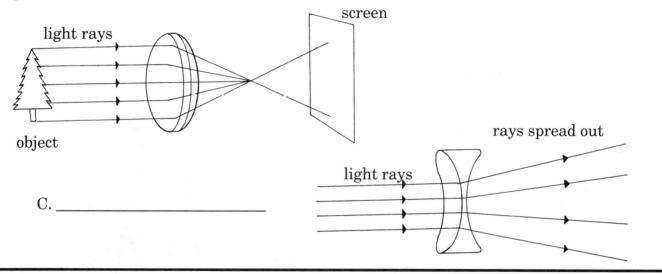

A. _____   B. _____

Image Formation

C. _____

1. What is a lens? _____
   _____

2. Label the type of lenses on lines A and B above.

3. Write the word or words that will make each sentence a true statement.

   a. A lens with two surfaces curving inward is called _____, while a lens with two surfaces curving outward is _____.

   b. Since the image in a camera is inverted, the camera has a _____ lens.

   c. The curvature of an eye's convex lens is changed by _____ attached to it.

   d. When a person's eye lens is too convex, images are focused in front of the retina and the person is _____-sighted.

   e. People who are _____-sighted see things well at a distance but not nearby.

4. On line C, draw the correct position of the image as it appears on the screen.

5. Put an X at the point where the rays meet as light passes through a convex lens. What is the name of this point? _____

6. What happens to light rays as they pass through a concave lens? _____
   _____

© Milliken Publishing Company — Light and Sound Energy

Optical Instruments

camera
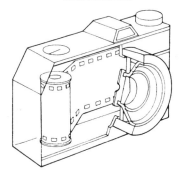

refracting telescope
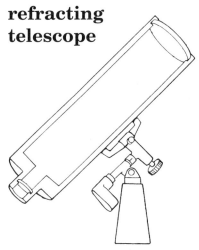

movie projector
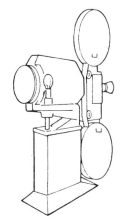

microscope
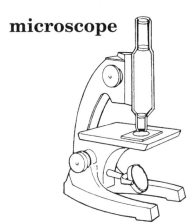

---

1. In each illustration above, circle the part that refracts light rays.
2. The most common type of lens used in optical instruments is a _____ lens.
3. Draw arrows to show the direction of light rays in each optical instrument shown above.
4. Write the word or words from Column B in the space before its description in Column A.

Column A	Column B
_____ a. used to examine small objects	refracting telescope
_____ b. used to view distant objects	movie projector
_____ c. records image on film	camera
_____ d. projects image on screen	microscope

5. How are most optical instruments focused? _____
   _____
6. Which instrument is most like the human eye? Why? _____
   _____

© Milliken Publishing Company — Light and Sound Energy

Television

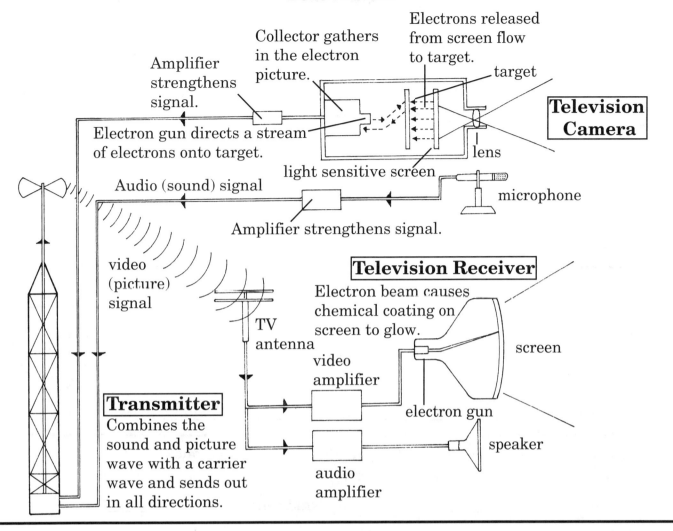

1. What are the three basic parts of a television system? _____

2. Identify the following parts of a television system:

    a. _____ shoots beam of electrons onto target plate

    b. _____ focuses light onto camera screen

    c. _____ collects returning electron beam

    d. _____ sends sound and picture waves

    e. _____ picks up video signal from atmosphere

3. What is the function of an amplifier? _____

4. What causes the image to appear on the television screen at home? _____

5. What is the function of a satellite dish antenna? _____

Sound

Sound waves are the result of vibrations.

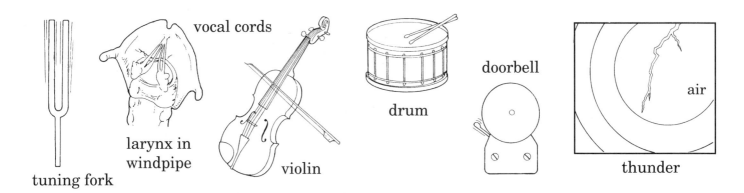

tuning fork • larynx in windpipe • vocal cords • violin • drum • doorbell • thunder / air

Vibrations

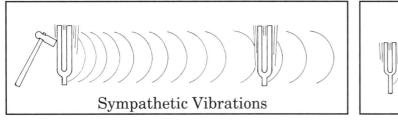
Sympathetic Vibrations

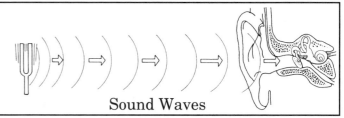
Sound Waves

1. In each picture above, place an X on the part that vibrates to produce sound.
2. Name the part on each object that vibrates to produce sound:

   a. church bell _____   d. wind chime _____

   b. radio _____   e. dog barking _____

   c. piano _____   f. guitar _____

3. Write the word or words that will make each sentence a true statement.

   a. Sound waves are the result of _____ in the air or other media.

   b. An object will no longer produce sound waves when it has stopped _____.

   c. Vibrating objects send out _____ _____ that can travel through solids, liquids, or gases but not through a _____ .

   d. The vibrations of the eardrum cause tiny _____ in the inner ear to vibrate.

   e. The _____ nerve relates sensation to the brain, which interprets the stimulus.

4. Describe how we hear the ringing of the bell. _____

5. What allows us to distinguish one sound from another when our brain receives vibrations?

© Milliken Publishing Company — Light and Sound Energy

Sound Waves

The **sound wave** is a series of compression and rarefaction areas traveling through a substance.

The Speed of Sound

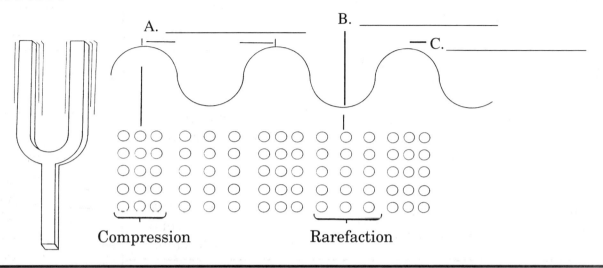

Compression Rarefaction

The Speed of Sound

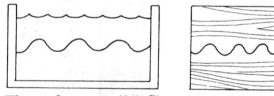

Through water (25°C) Through pine wood (25°C) Through steel (25°C)
1497 m/sec 3320 m/sec 5200 m/sec

> The speed of sound is 331.5 meters per second in air at 0°C. The speed increases about 0.6 meters per second for each degree rise in temperature (C).

---

1. Label the parts of a sound wave on lines A–C above.
2. Describe the relative spacing of molecules as they are affected by compression and rarefaction. _____

3. Describe the motion of individual molecules. _____

4. What is the speed of sound through air at these temperatures? (Remember to use m/second.)

   a. 0° C _____   c. 25° C _____   e. 38° C _____   g. 56° C _____

   b. 20° C _____   d. 30° C _____   f. 42° C _____   h. 68° C _____

5. Look at the speed of sound through these unknown substances and determine if the substances are solid, liquid, or gas.

   a. 1200 m/sec _____   c. 5000 m/sec _____

   b. 259 m/sec _____   d. 2680 m/sec _____

© Milliken Publishing Company   Light and Sound Energy

Characteristics of Sound

I. Frequency— _____

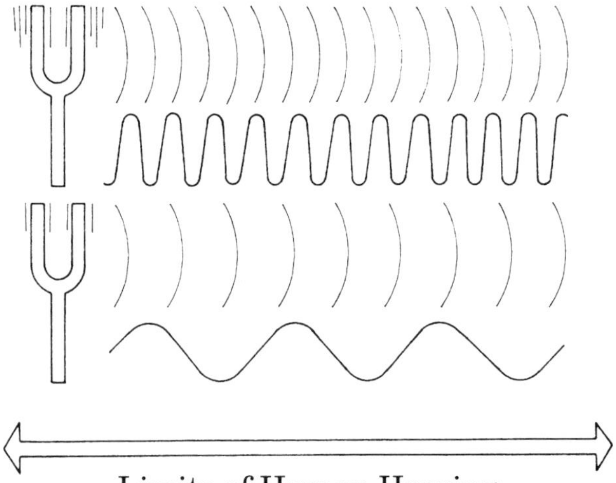

II. Amplitude— _____

III. Intensity— _____

IV. Loudness— _____

16 vibrations per second (*subsonic*) ← Limits of Human Hearing → 16,000 vibrations per second (*ultrasonic*)

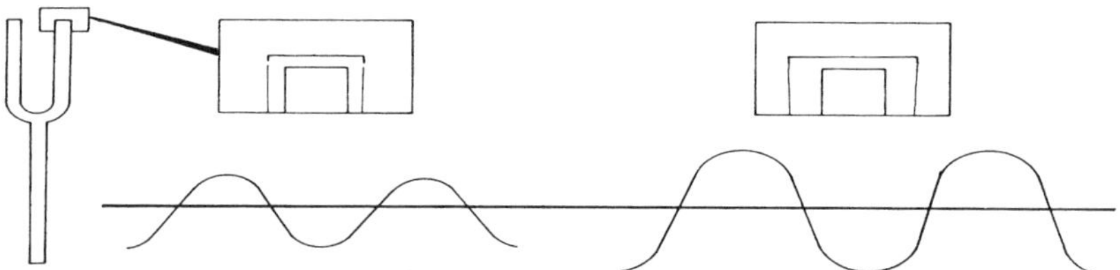

small amplitude— small sound
large amplitude— large sound

Decibels
- 0 silence
- 10 heartbeat
- 20 whisper
- 30 soft piano
- 40 typewriter
- 50 soft talking
- 60 conversation
- 70 barking dog
- 80 street traffic
- 90 food blender
- 100 air hammer
- 110 thunder
- 120 painful sounds
- 170 jet engine

1. On lines I through IV, write the definitions for frequency, amplitude, intensity, and loudness.
2. Is the frequency of the following sounds high or low?

   a. whine of jet engine _____ c. typewriter _____ e. thunder _____

   b. whisper _____ d. air hammer _____ f. heartbeat _____

3. Write the word or words that will make each sentence a true statement.

   a. A sound with many vibrations per second (vps) produces a _____ pitch; a sound with few vps produces a _____ pitch.

   b. Whistles heard by dogs and not people have more than _____ vps.

   c. Sounds of fewer than 16 vps cannot be heard by people and are said to be _____.

   d. Loudness and intensity are not the same concept due to the interaction of _____.

   e. Sounds above _____ decibels are painful to the ear and should be avoided.

4. Try to guess how many decibels the following sounds produce:

   a. rock concert _____ c. firing a large cannon _____

   b. humming a song _____ d. gentle breeze through trees _____

© Milliken Publishing Company — Light and Sound Energy

Sound and Materials

Sound waves travel best in materials when the molecules are close together.

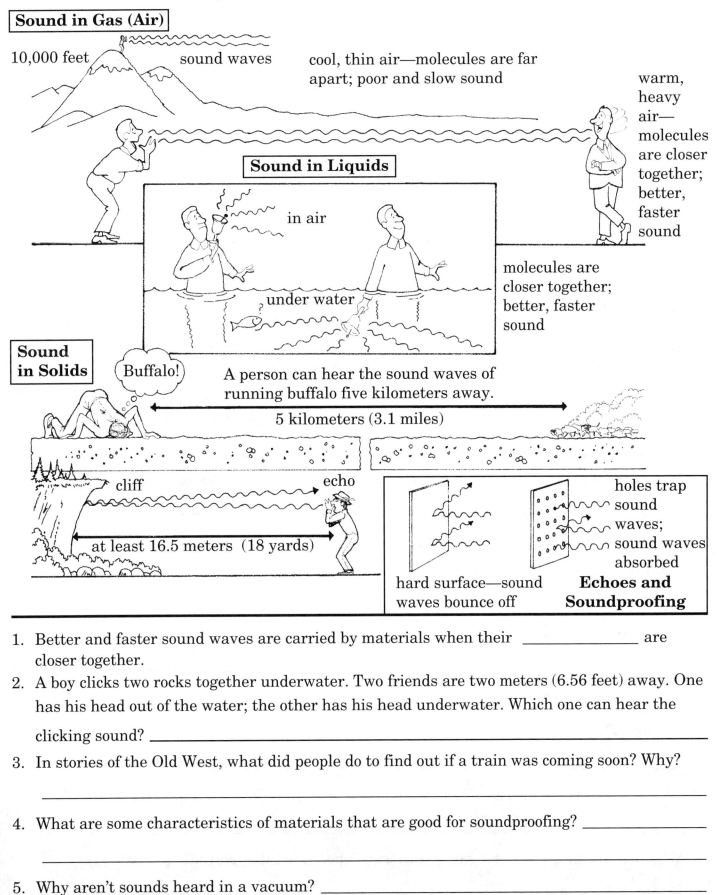

1. Better and faster sound waves are carried by materials when their _____ are closer together.

2. A boy clicks two rocks together underwater. Two friends are two meters (6.56 feet) away. One has his head out of the water; the other has his head underwater. Which one can hear the clicking sound? _____

3. In stories of the Old West, what did people do to find out if a train was coming soon? Why?
_____

4. What are some characteristics of materials that are good for soundproofing? _____

5. Why aren't sounds heard in a vacuum? _____

© Milliken Publishing Company   10b   Light and Sound Energy

Musical Instruments

String instruments

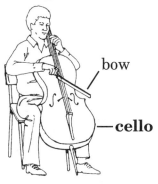
bow
cello
Stroke strings to start vibration.

banjo
Pluck strings to start vibrations.

Wind instruments

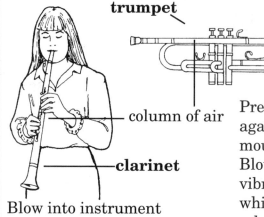
trumpet
column of air
clarinet
Blow into instrument to vibrate column.
Press lips against mouthpiece. Blow air to vibrate lips which vibrate column.

Percussion instruments

Strike with metal bar. Solid metal vibrates when struck.

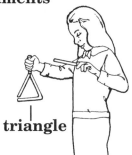
triangle

drum
stretched material
hollow container filled with air
Strike with stick.

1. What part of these instruments vibrates to produce sound? What causes this part to vibrate?

   cello _____

   clarinet _____

2. List three ways to raise the pitch of a string instrument. _____
   _____

3. How would you raise or lower the pitch of a wind instrument? _____
   _____

4. How can the pitch be changed in a triangle? _____

   in a drum? _____
   _____

5. Why is a piano classified as a percussion instrument even though it has many strings?
   _____

6. What causes the strings of a guitar to vibrate when it is being played?
   _____

© Milliken Publishing Company   Light and Sound Energy

The Human Voice

Overhead View of Vocal Cords

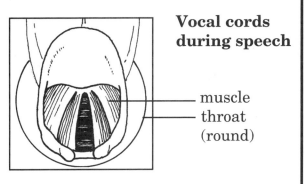

Vocal cords during speech — muscle, throat (round)

Vocal cords at rest

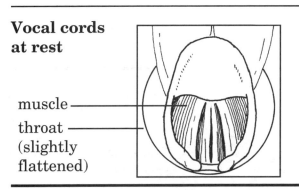

muscle, throat (slightly flattened)

Parts of Voice Box

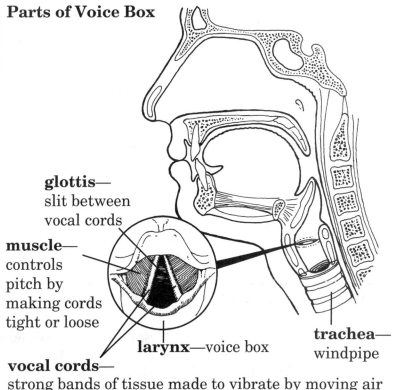

- **glottis**—slit between vocal cords
- **muscle**—controls pitch by making cords tight or loose
- **larynx**—voice box
- **trachea**—windpipe
- **vocal cords**—strong bands of tissue made to vibrate by moving air

Human Vocal Cords

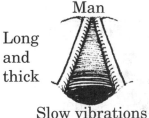

Man — Long and thick — Slow vibrations

Woman — Thin and short — Fast vibrations

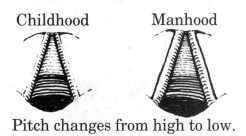

Childhood — Manhood — Pitch changes from high to low.

1. The opening between the vocal cords is called the _____; vocal cords vibrate when _____ moves rapidly between them.

2. Label the following activities with a V if the vocal cords are used and with NV if the vocal cords are not used:

   a. singing _____   c. humming _____   e. whistling _____

   b. sleeping _____ d. thinking _____   f. crying _____

3. How is pitch controlled in the human voice? _____
   _____

4. a. What is another name for the trachea? _____

   b. Why is this an appropriate name? _____

5. The normal pitch of a woman's voice is *(high, low)*; the normal pitch of a man's voice is *(high, low)*.

© Milliken Publishing Company — 11b — Light and Sound Energy

Development of Communication Systems

Sending Long Distance Messages

Sending Messages with Electricity

sender or receiver

Person receiving signal changes dot and dash sounds into the alphabet.

sender or receiver

Sending Messages with Voice and Electricity

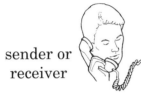

sender or receiver

Sound energy changed to electrical energy.

sender or receiver

Communicating with Many People

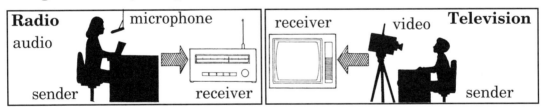

1. To communicate, a _____ and a _____ are needed.

2. What were the advantages of drum signals over smoke signals? _____
   _____

3. What were the disadvantages of the telegraph for communication? _____
   _____

4. The sound part of a television program is called the _____ portion; the picture part is the _____ portion.

5. In radio communication, _____ energy is changed to _____ energy and then to _____ energy.

6. List three inventions that use a person's voice and electrical energy to transmit sound.
   _____

© Milliken Publishing Company   12a   Light and Sound Energy

A Last Look—Part I

A. In each of the following groups one item does not belong. Circle that item and in the space provided explain why it does not belong.

1. x–rays	radio waves	sound waves
2. compression	refraction	reflection
3. moonlight	firelight	sunlight
4. red	black	blue
5. convex	screen	concave
6. radio	microscope	camera
7. crest	photon	trough
8. violin	piano	telephone
9. pitch	particle	quantum
10. amplitude	frequency	rarefaction

B. Write the word or words that will make each sentence a true statement.

1. Light rays and infrared rays are kinds of _____ energy.
2. Luminous objects produce _____.
3. A wavelength is measured from crest to _____.
4. The speed of light is about _____.
5. Light passing through a _____ produces a spectrum.
6. _____ materials in a beam of light produce a shadow.
7. Sound waves are the result of _____.
8. Sound travels faster in a _____ than in a gas.
9. Vibrations of more than 16,000 per second are called _____.
10. Moving air causes our _____ to vibrate and we hear sounds.

© Milliken Publishing Company   I   Light and Sound Energy

A Last Look—Part II

A. On line a, name the instrument shown. On line b, tell what type of lens is in the instrument.

 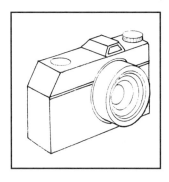  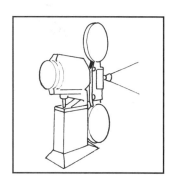

1a. _____ 2a. _____ 3a. _____ 4a. _____

b. _____ b. _____ b. _____ b. _____

B. On line a, name the instrument shown. On line b, tell how the pitch is changed.

   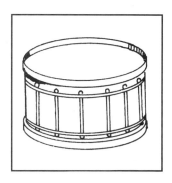

1a. _____ 2a. _____ 3a. _____ 4a. _____

b. _____ b. _____ b. _____ b. _____

C. On line a, name the object shown. On line b, tell if it is luminous or illuminated.

 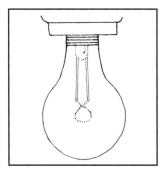  

1a. _____ 2a. _____ 3a. _____ 4a. _____

b. _____ b. _____ b. _____ b. _____

© Milliken Publishing Company Light and Sound Energy

A Last Look—Part III

A. Find the statement in the second column that best describes each word in the first column. Write the letter of the statement before the word it describes.

1. _____ ultraviolet rays
2. _____ photon
3. _____ refraction
4. _____ rainbow
5. _____ translucent
6. _____ concave
7. _____ eyepiece
8. _____ rarefaction
9. _____ amplitude
10. _____ telegraph

a. bundle (quantum) of energy
b. molecules are farther apart
c. type of camera
d. bending of light rays
e. communication device
f. magnifies image
g. displacement of vibrating particle
h. spectrum of refracted light
i. spreads out light rays
j. angle of incidence
k. image is diffused
l. invisible radiant energy

B. Circle the word that will make each sentence a true statement.

1. The _____ is a communication device that uses both an audio and video signal.
   telegraph   television   radio
2. Speech is produced by the _____ located in the larynx.
   throat muscles   vocal cords   trachea cartilage
3. Tightening the strings on a banjo produces a higher _____ .
   echo   amplitude   pitch
4. Sounds travel more slowly in _____ because the molecules are farther apart.
   solids   liquids   gases
5. Loudness of sounds is measured in _____ .
   decibels   photons   refractions
6. When a nearby tuning fork vibrates without being struck, it is an example of _____ vibrations.
   compression   sympathetic   amplitude
7. _____ lenses are found in most optical instruments.
   Concave   Mirror   Convex
8. Wood is an example of a(n) _____ material.
   translucent   opaque   transparent
9. When all colors of the spectrum are reflected by an object, it appears _____ .
   white   black   red
10. Light rays _____ when they pass from air into glass.
    speed up   slow down   do not change

© Milliken Publishing Company — Light and Sound Energy

A Last Look—Part IV

A. Explain fully the meaning of this cartoon.
_____
_____
_____
_____
_____
_____

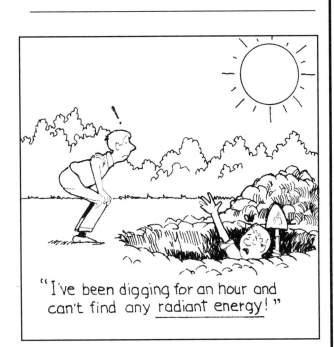

"I've been digging for an hour and can't find any radiant energy!"

B. There is something wrong with each of these drawings. Circle the part of the drawing that is incorrect and explain why you circled it.

1. _____   2. _____
   _____   _____

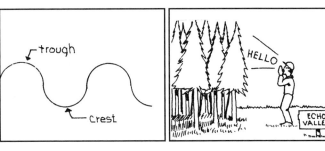

3. _____   4. _____
   _____   _____

C. Use the clues to complete the puzzle. When you have finished, the letters in the dark squares will spell a word that describes a part of a sound wave.

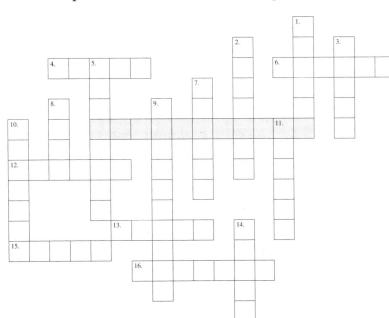

1. bundle of light energy
2. unit of energy
3. what we hear
4. produces spectrum
5. radiant energy producing heat
6. produced by a prism
7. caused by opaque objects
8. percussion instrument
9. produced by a mirror
10. sound measurement
11. not transparent
12. optical instrument
13. produced by high and low frequency
14. light and sound travel in _____
15. enables you to see
16. kind of lens

Light and Sound Energy

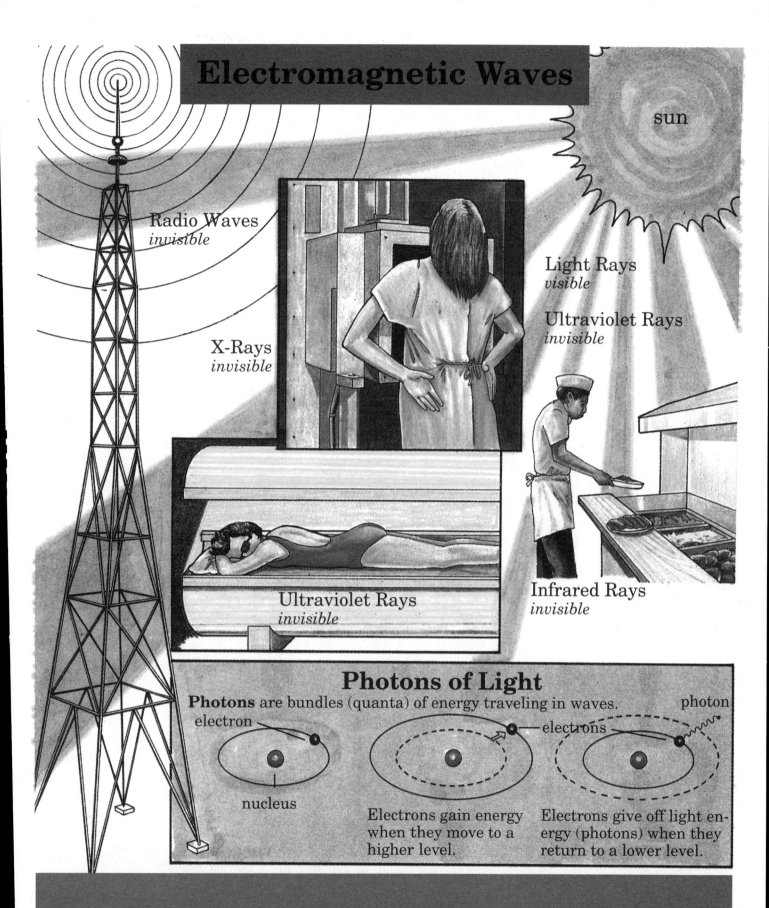

1. Which form of electromagnetic wave is visible to the human eye?
2. What causes an atom to give off light?

STUDY QUESTION: Find out about the electromagnetic spectrum.

Light Waves

Theories of Light

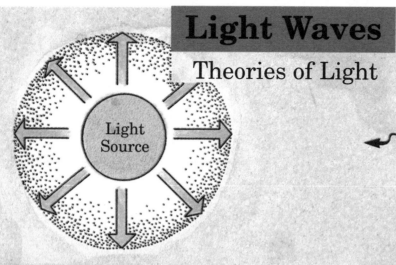

Particle Theory of Light
Streams of tiny particles move in all directions from the source, like buckshot.

Wave Theory of Light
Light travels in waves in all directions from source, like waves from a disturbance in water.

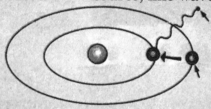

Quantum Theory of Light
Bundles of light energy, called **quanta,** travel in waves.

Parts of a Light Wave

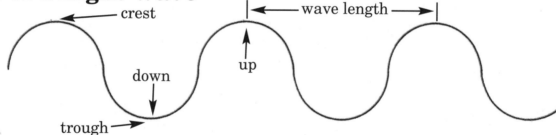

Speed and Movement of Light

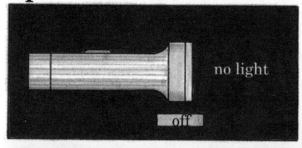
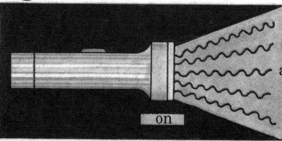

Light waves travel in a straight line at a speed of about 300,000 km/sec away from the source.

1. How is the quantum theory of light a combination of the other two theories?
2. Describe the movement of light.

STUDY QUESTION: Research the visible portion of the electromagnetic spectrum.

Reflection and Refraction

Reflection of Light

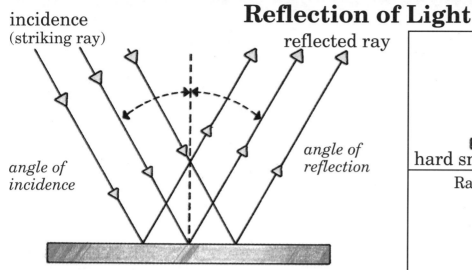

The **Law of Reflection** says that the angle of incidence is equal to the angle of reflection.

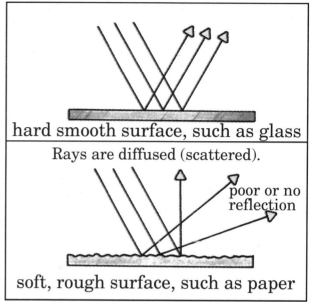

Refraction of Light

Light rays bend as they pass through a glass lens.

Depending on the density of the transparent material, the speed of light will change, causing refraction.

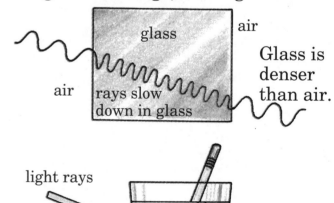

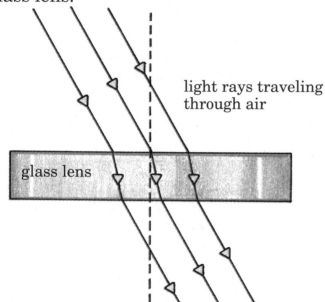

Refraction is the bending of light rays.

An Example of Refraction

1. Why might there be no reflection from a rough surface?
2. What causes refraction?

STUDY QUESTION: What causes mirages?

Light and Color

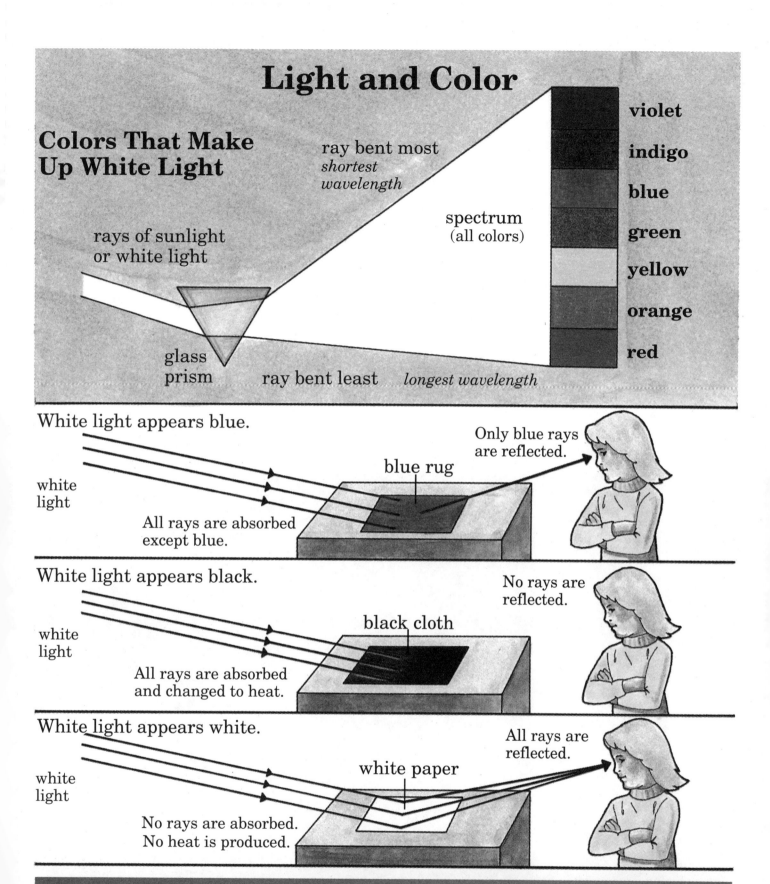

1. Name the colors of the spectrum, starting with the color that has the longest wavelength and ending with the color that has the shortest wavelength.
2. Why does a rose appear red to our eyes?

STUDY QUESTION: Why is outer space black?

Light and Materials

Kinds of Material and Light

Transparent Materials

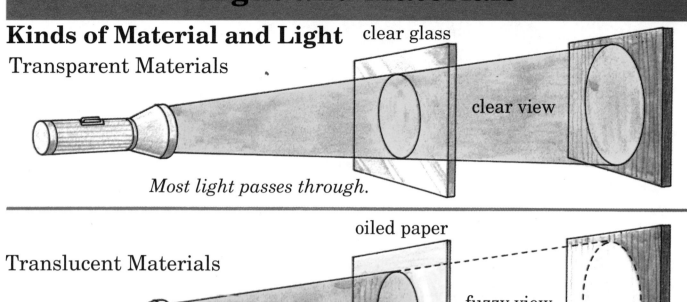

Most light passes through.

Translucent Materials

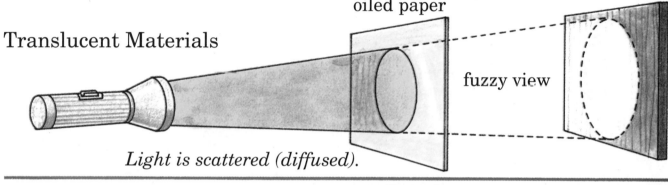

Light is scattered (diffused).

Opaque Materials

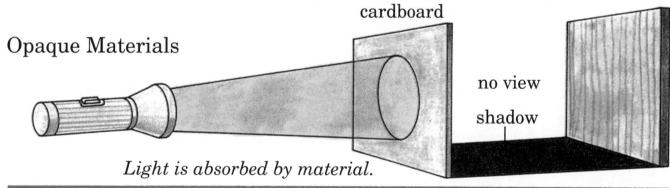

Light is absorbed by material.

Light and Shadows

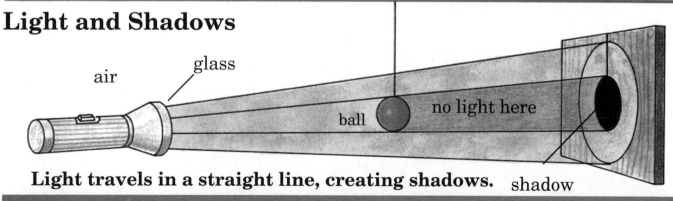

Light travels in a straight line, creating shadows.

1. Which type of material allows a person looking through to see objects on the other side most clearly?
2. What causes a shadow?

STUDY QUESTION: Find out about the umbra and penumbra parts of a shadow.

Optical Instruments

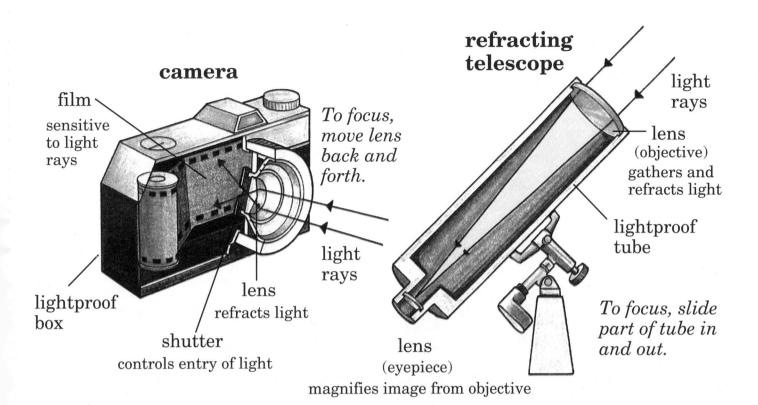

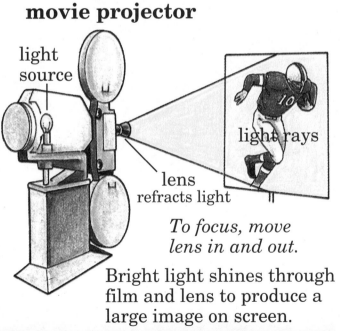

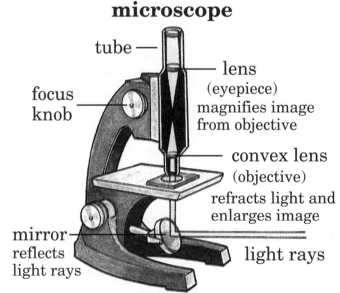

1. What type of lens is found in most optical instruments?
2. What are the functions of the two lenses in a refracting microscope?

STUDY QUESTION: Learn about the lenses and prisms in binoculars.

Sound

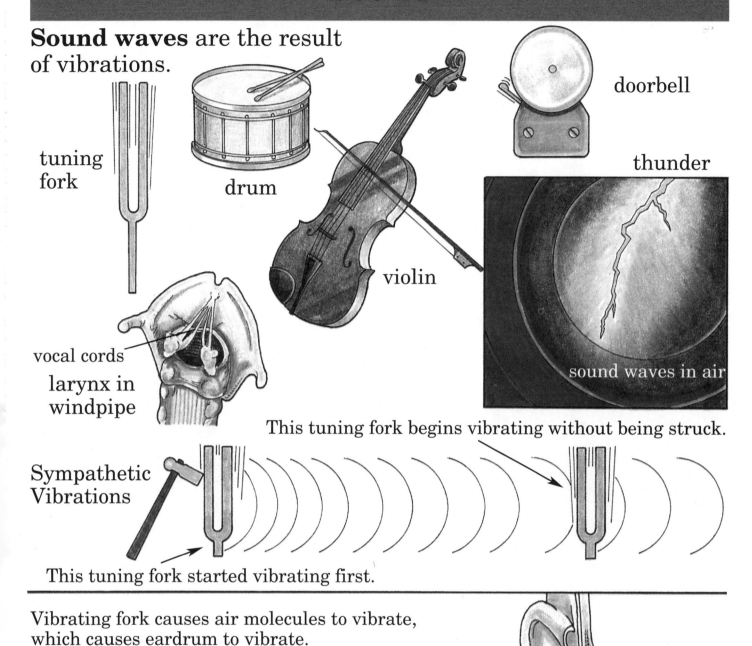

1. What causes sound?
2. How do we hear sound?

STUDY QUESTION: Find out about the sound of space vehicles on the surface of the moon.

Sound Waves

The **sound wave** is a series of compression and rarefaction areas traveling through a substance.

Compression
When the tuning fork moves forward, it compresses the air so that molecules are close together.

Rarefaction
When the tuning fork moves back, the air is rarefied so that the molecules move farther apart.

Individual molecules do not travel but vibrate rhythmically back and forth.

tuning fork

The Speed of Sound

The speed of sound is 331.5 meters per second in air at 0°C. The speed increases about 0.6 meters per second for each degree rise in temperature (°C).

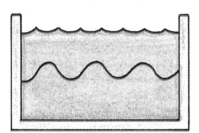

through water (25°C)
1497 m/sec

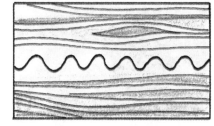

through pine wood (25°C)
3320 m/sec

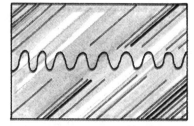

through steel (25°C)
5200 m/sec

1. Describe how a sound wave travels.
2. Does sound travel fastest through a solid, a liquid, or a gas?

STUDY QUESTION: Find out how an echo is produced.

Characteristics of Sound

Frequency is the number of vibrations per second.

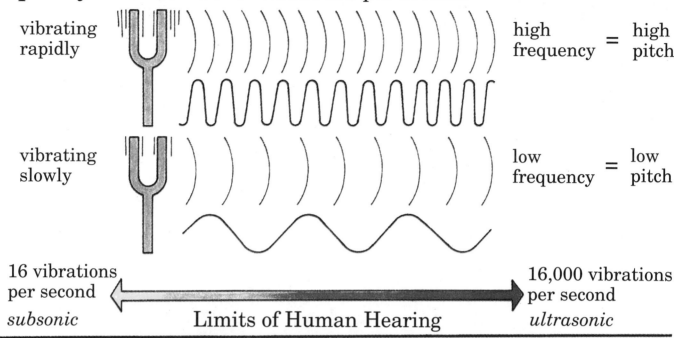

vibrating rapidly — high frequency = high pitch

vibrating slowly — low frequency = low pitch

16 vibrations per second — *subsonic* ←——— Limits of Human Hearing ———→ 16,000 vibrations per second — *ultrasonic*

Amplitude is the distance a vibrating particle has been displaced from a resting position.

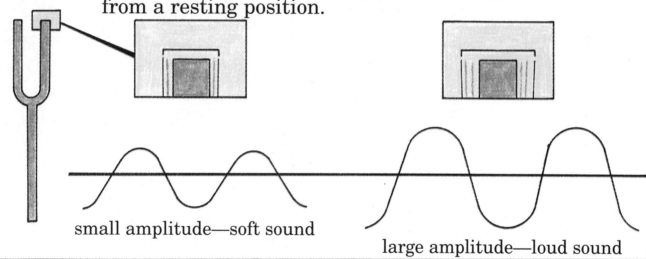

small amplitude—soft sound

large amplitude—loud sound

Loudness is the strength of sounds, measured in decibels.

silence	whisper		type-writer		conver-sation		street traffic		air hammer		painful sounds				jet engine
0	10	20	30	40	50	60	70	80	90	100	110	120			170
	heartbeat		soft piano		soft talking		barking dog		food blender		thunder				

1. What kind of sound would be produced by a rapidly vibrating object?
2. Can you hear a sound produced by 20,000 vibrations per second?

STUDY QUESTION: Find out about the Doppler effect.

Musical Instruments

String instruments use one or more vibrating strings to produce sounds.

Cello
bow

Stroke strings to start vibration.

Banjo
hand

Pluck strings to start vibration.

Pitch Change

higher
- tight strings
- short strings
- thin strings

lower
- loose strings
- long strings
- thick strings

Wind instruments use a column of vibrating air to produce sound.

Pitch Change

higher
- short air columns

lower
- long air columns

Clarinet

column of air

mouth-piece with vibrating reed

Blow into instrument to vibrate column.

Trumpet

Press lips against mouthpiece. Blow air to vibrate lips, which vibrates column.

Percussion instruments, when struck, produce sounds by vibrating.

Drum

stretched material

air–filled container

Strike with stick.

Pitch Change

Drum:
- Tighten or loosen cover.
- Use thick or thin material.
- Use large or small piece for cover.

Triangle:
- Use different material.
- Change length of material.

Triangle

Strike with metal bar.

Solid metal vibrates when struck.

---

1. How is the pitch changed on a stringed instrument?
2. Why does a bass drum have a low pitch?

STUDY QUESTION: Find out how these instruments are classified: oboe, harp, celesta, glockenspiel, and zither.

Development of Communications Systems

Sending Messages with Sound

Sending Messages with Electricity

Person receiving signal changes dot and dash sounds into the alphabet.

Sending Messages with Voice and Electricity

Sound energy changes to electrical energy.

Communicating with Many People

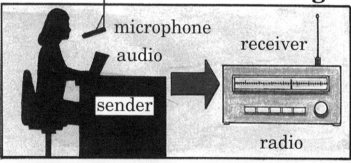

1. What energy source is most often used for communications?
2. How are telephone and radio communication different?

STUDY QUESTION: Research various communication systems used by animals.

© Milliken Publishing Company   12   Light and Sound